The Trouble With Zero

AJ Corcoran

The Trouble With Zero

ISBN   978-0-646-53723-8

www.thetroublewithzero.net

Published by Palo Pacific Technology Pty Ltd, Sydney, Australia

palo pacific technology

Rev D

The Trouble With Zero

Table of Contents

List of Figures

1 Introduction

Would you buy a car that only works some of the time?

Would you accept a theory of gravity that only works some of the time?

Why accept a theory of zero that only works some of the time?

This is a book about zero. This is not about the zero you learnt about in school. This is a tale of three different types of zeros, a misunderstanding and the impact it has had on Western mathematics and physics. It is one of the strangest tales in the history of mathematics and science.

It is no coincidence that the most controversial areas of mathematics are also involved in many of the unexplained problems in physics. After thousands of years of research, you would think that we would know all there is to know about basic algebra. However, as our understanding of mathematics and physics has advanced, it has exposed deficiencies in our theories about numbers and basic mathematics.

These deficiencies show up as strange or undefined results in the mathematics of physics. Our misunderstandings lead to strange conclusions, such as that the laws of physics must break down sometimes or that the mathematics of the Universe does impossible things.

It is generally assumed that this is "just the way things are" and that the Universe has some strange behaviour. The actual truth is much simpler: human error. A few bad assumptions made centuries ago still colour our views and lead us to the wrong conclusions.

Much of what we learn about zero in school is wrong. It is based on opinion and philosophy dating back many centuries, rather than being based on the way that the Universe actually does things.

There are several different types of zero in common use. Each of these uses started at a different time and for different reasons. Confusion about these different types of zeros is at the heart of many controversies involving zero.

In this book we will examine the properties of three types of zeros as well as where and how this confusion arises and the effect it has on mathematics, geometry and physics.

2 The Beginnings of Zero

Although this is not a history book, we cannot fully understand our concept of zero without understanding how we arrived at it. Zero as we know it today did not pop out as a fully-formed idea, but developed over time.

A Very Brief History of Zero

Early man knew that he either had some food or no food–something or nothing. Many thousands of years ago, our ancestors could distinguish between zero and non-zero numbers. Zero has passed through many hands on its way from its use as a physical quantity, to its use as a place-holder in the Babylonian numbering system, to its use as a number in its own right. Our modern concept of zero has come to us from our distant ancestors, via Babylon, Egypt, Greece, India and Arabia.

The introduction of zero as we now know it was not an easy one and came with considerable controversy. Much of this revolved around whether zero was really a number of not. The side that said that zero was a real number defeated those who said it was not. As we shall see, in actual fact they were both right or both wrong, depending on how you look at it.

The Babylonian Placeholder

It is generally agreed that the Babylonians were the first to use zero. The way they used it though was quite different to the way we do. They used it more as a punctuation mark than as a number and they did not use it to mean "nothing". The Babylonians were among the first to use a "place value" number system. They used a number system based on the number 60. This is called a sexagesimal system. Their script is called cuneiform from Latin "cuneus", a wedge.

Babylonians first used a form of zero as a place-holder in around 300 BCE. They used it much as we do today when writing something like "101", to denote that a value (in this example, the tens) was missing. They had a similar problem with their number system that we have with ours; to write one hundred and one without using zero we end up with "1 1". This could also be interpreted as eleven. To eliminate any confusion a zero place-holder can be used to separate the units from the hundreds.

The Babylonians treated their place-holder as an operator or a punctuation mark rather than as a number. It was just there to separate the columns that contained the actual numbers. The "placeholder zero" was used to denote the absence of a value in a sequence, not an actual number.

Other Uses of Zero as a Placeholder

This use of zero as a placeholder to mean "there are none of these" was developed independently in South America, Central America and India, and other places. The use of zero in early placeholder systems was generally more like a punctuation mark than how we use it today.

A zero place-holder was also used in China, although they didn't adopt zero as a number until it was introduced from India by a Buddhist astronomer in 718 CE. It seems that this original transition from "zero the placeholder" to "number zero" happened in India and spread west to Europe and east to China.

All these systems are remarkably like our own base-ten decimal system. We use the digit "0" to denote a column in our number that is not included in the total.

This use as a placeholder led to the use of zero as a number in itself. The thinking was that *if zero could be used as a numeral then it must be a number...* This was one of the causes of the trouble we are examining here.

Over the years we have invented new ways of using zero. For example, we still use it as a placeholder between 9 and 1 in the decimal number system, as in: 19, 20, 21. We also use it to fill a place between 59 and 1 when counting seconds or minutes, as in 1:59, 2:00, 2:01. Note that this is purely a man-made concept and has nothing to do with the way that nature treats time.

Indian Mathematicians

The first uses of zero as we know it today are credited to Indian mathematicians such as the authors of a Jaina text entitled the Lokavibhaaga in 458 CE. The use of a blank space to represent a zero dated back to at least the 4th century BCE.

India had a strong tradition of mathematics over many centuries. Today, the best known Indian mathematician of this time is Brahmagupta, who lived during one of the turning points in Indian mathematics.

Brahmagupta

The Indian mathematician and astronomer Brahmagupta (ca 598–665 CE) wrote about zero as a number in Brahmasphuta-siddhanta (The Opening of the Universe) in 628 CE. This is often described as the "golden age" of Indian mathematics. Brahmagupta is often referred to as the inventor of zero due to his writings on the subject, although he was almost certainly influenced by earlier Indian mathematicians.

Brahmagupta wrote down many of the rules that we take for granted these days: how to add, subtract, multiply and divide, as well as how to solve quadratic equations. Although Brahmagupta treated zero as a number he still wrote that $0/0 = 0$, as in the traditional use of "nothing". It was only much later that division by zero was treated as undefined.

Cantor, Infinity and Set Theory

Georg Cantor (1845–1918) is well known for his work on countable and uncountable infinities. His other great contribution was in set theory, which dates from 1873.

Although it is rarely recognised, set theory was also a very great advancement for zero. It allows us to distinguish between "Nothing" or the empty set { } and "Number Zero" { 0 } or the set with a single number in it. Without the ability to make this distinction, any discussion on whether zero is a number or not ends up being a matter of opinion or philosophy, rather than mathematics.

3 The Different Types of Zeros

Zero is not just a single thing.
We have several kinds of zeros in everyday use.
We frequently apply the rules of one kind of zero inappropriately to another kind.

Zero is not a single entity. There are several types of zero in common use, including:

- As a placeholder in our numbering system.
- As an ordinal number
- As a quantity

It is generally recognised that there is a difference between using zero as a placeholder (as in "101") and as an actual number (as in "I have 0 apples"). It is not often recognised that there is also a difference between using zero as an ordinal value and as a quantity. In this book we will examine each of these uses of zero.

Placeholders, Ordinals and Quantities

We commonly divide zeros into placeholders (such as the 0 in 102) and numbers (such as the 0 in the series -1, 0, 1, 2). However, there is a further distinction that is often overlooked: the use of zero as an ordinal value as opposed to its use as a quantity. "Ordinal numbers" show rank, order or position, not quantity. Numbers used to describe quantities are also referred to as "cardinal numbers".

Ordinal values are a human invention. The minutes on the clock, powers of 10, array indices are all human inventions. The mathematics of nature cares little for integers, clocks and indices.

For example, "zero time" 0:00, meaning no hours or minutes elapsed is a natural concept, but the "zero hour" between 23:00 and 1:00 is a man-made construct.

Similarly we could mark a 10 centimetre ruler thus:

0	1	2	3	4	5	6	7	8	9

But, once again, the 0th section is 1cm long not zero-length. This is another example of the difference between zero used as an ordinal number and as a physical quantity.

Zero as a Placeholder

The use of zero as a placeholder, for example the zero in "101", has been around since Babylonian times. In the case of "101", zero is used to indicate that there are no tens in the number.

It also helps us differentiate "101" from "1 1", which we would have to write if we did not have a numeral for zero. If we wrote "101" in base 20 it would be "51" and the zero would not appear. This use of zero as a placeholder is quite distinct from a "real" zero value.

Zero as an Ordinal Number

We can use zero as an integer for the purposes of counting. This is the case when we use zero as an ordinal value.

Examples where zero is an ordinal value:
- Powers of two: 2^{-1}, 2^{0}, 2^{1}
- Latitude, longitude, other coordinates
- Computer software pointers
- Array indices
- Hours : minutes : seconds on a digital clock

Note that these are "man-made" uses of zero, as opposed to the way that physical quantities appear in the mathematics of Physics.

Zero as a Quantity

Zero may also represent a quantity, as opposed to being a placeholder or an ordinal value. Let's take the example of apples. If we have no apples then we have the set of { } apples – our set is empty and we have a "null quantity".

Some examples where zero may be a null quantity:
- Mass
- Energy
- Time
- Distance
- Money
- Apples
- Cats

Zero as Both an Ordinal and a Quantity

We sometimes use zero as both a quantity and an ordinal value in the same expression.

Example 1

Zero is used as both a quantity and an ordinal number in:

0^0 (0^0 or zero raised to the 0th power).

Example 2

In the number 10.01 the placeholder zeroes provide ordinal integer values between 9 and 1, but they also represent an absence of value (no units or tenths).

Zero and the Number Line

Here's an example of the different ways we use zero on the number line. Below is a chart showing the number of cats in my street over the course of 24 hours.

Figure 1   Cats In My Street

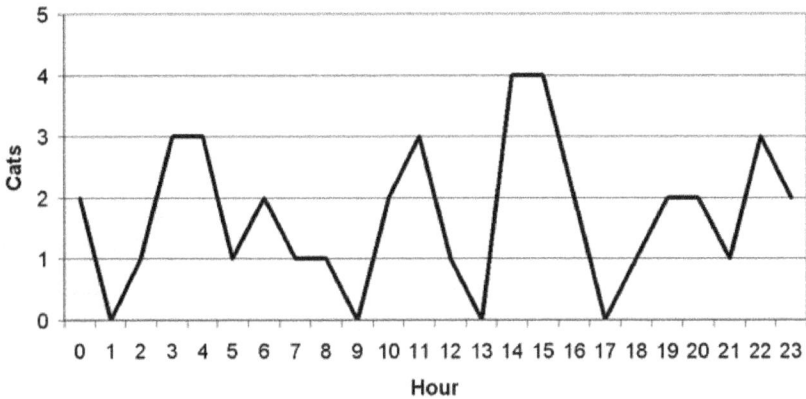

Note that the chart contains three different types of zeros:

- As a placeholder in hours "10" and "20".
- As a quantity in the number of cats.
- As an ordinal value as "hour 0".

Figure 2   Three Types of Zeros

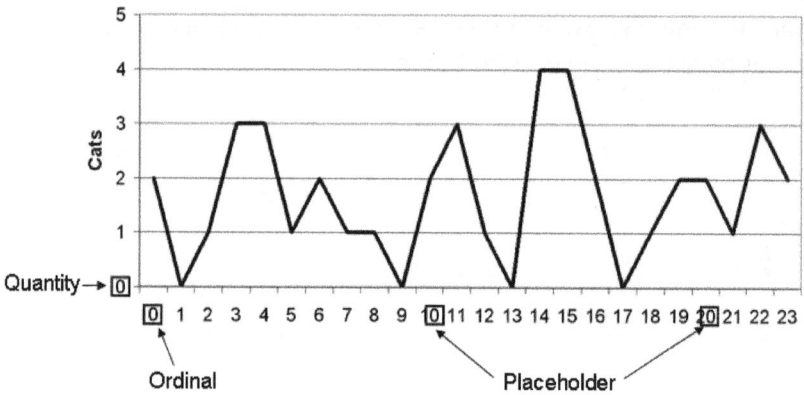

Notice that the ordinal hour "0" on the X axis has the same width as hour 1 or hour 2. Similarly, the hours "10" and "20" occupy an hour. On the Y axis however, a zero quantity has no height–it occupies no length at all on the axis. No cats at all may be thought of as the absence of any cats.

We take it for granted that zero is always a number. We have been trained to do so since childhood. However, as we observe on the number line, there are times we use it as a one-dimensional value and others when we use it as a zero-dimensional value.

The rules for 0- and 1-dimensional values are not exactly the same, just as the rules for 2-dimensional values, such as (x, y) coordinates, are not exactly the same as for a 1-dimensional value.

How Wide Is Zero?

We note that zero as a quantity is often shown on number lines as occupying some width on the line. We draw actual lines to represent zero on our chart. This reinforces the convention that zero is an actual number. But how wide is zero as a quantity? What dimensions does it have on a number line?

In the chart "Cats In My Street" we chart a quantity (the number of cats) against ordinal values (the hours 0 to 23). This example demonstrates the difference between the use of zero as a quantity and as an ordinal number.

Figure 3   Quantities and Ordinals

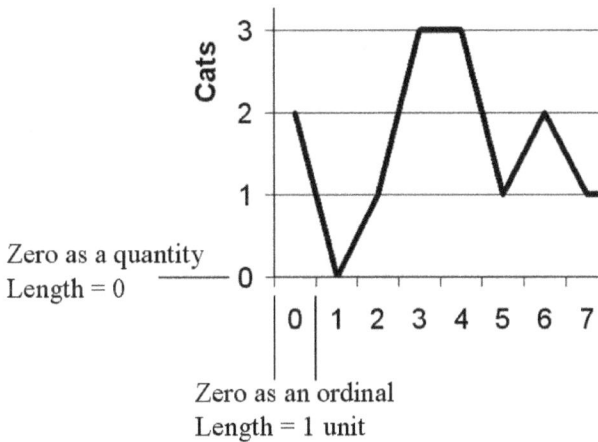

Observe that the zero on the vertical axis has no length, depth or height: it is 0-dimensional. Now observe that the zero on the horizontal axis has a length, although no height or depth: it is 1-dimensional.

The difference between these two types of zero is at the root of many of our problems with zero. It is our convention to treat both of these as 1-dimensional numbers, the same as one or two or any other number.

As a quantity, zero occupies no width on the number line. Even the slightest movement in the positive or negative directions results in a non-zero. It must always be remembered that as a physical quantity zero has zero height, zero width and zero length. Any other representation, such as a dot or a line, is a human invention.

Charts like the one above were not in common use until after the invention of Cartesian coordinates in the 17th century by René Descartes. For much of the history of zero, we have lacked simple tools such as this chart that can graphically show us the different types of zero sitting side by side.

When viewing zero as nothing, null or an empty set, we are not "taking zero off the number line" as it was never there to begin with. It is merely the zero-sized point that separates positive numbers from negative numbers. When used as a quantity, zero has no dimensions on the number line, as the slightest size in either direction would make it either positive or negative; in either case something other than zero.

"Nothing" versus "Number Zero"

Many thousands of years ago, man had the concept of "nothing". He either had something or he had nothing. Over the past thousand years though, we introduced the concept of "number zero". The introduction of zero was accompanied by often ferocious arguments. Along the way, the concept of nothing was merged into the concept of an integer half-way between -1 and 1. The result was "number zero" which is now used to describe both concepts.

Of these two ideas: "nothing" and "number zero", one is naturally-occurring (nothing) and the other is man-made (ordinal values).

"Nothing" is not the same as "number zero". The difference is best illustrated using set theory. The conventional view of number zero treats it as { 0 }, whereas nothing is { } or the empty set. In computing, this absence of any value is called "Null" or "Nothing".

You may sometimes see zero described as the empty set { } but immediately followed by its use as a number, or { 0 }. It is important to note that an empty set contains no information and its use as a number is incorrect. An empty set contains no elements, so there is nothing to operate on.

The difference between "number zero" and "nothing" can be shown by these two operations on a pocket calculator:

Operation A: One Divided by "Number Zero":
  Type "1 / 0 ="
  You'll get <Divide by Zero!> (or something similar.)

Operation B: One Divided by "Nothing":
  Type "1 /" without entering the "0" or "=" buttons.

You'll see the operation on an empty set.

Operation A insists on treating zero as an actual number and demanding a result.

Operation B is "incomplete" and there is no result as there is nothing to divide by.

Is Zero a Number?

Nothing highlights the problems we have created for ourselves with zero more than this simple question "Is Zero a Number?" Why? Because the answer is both yes and no–there are different kinds of zero in common use.

Nature does not treat zero as a number when it is a quantity–it treats it as "nothing" or the absence of a number, but human beings have invented "number zero" as an ordinal value and placeholder and introduced it into everyday use.

For example, "zero time" 00:00 is a natural concept, but the "zero hour" 00:00 between 23:00 and 1:00 is a man-made construct. Similarly the number 101 contains a zero, but it means that there are no tens in the number, not that the number itself is zero.

The expression 0^0 ($0\wedge0$, zero raised to the zeroth power) contains both an empty quantity and the man-made construct "to the 0th power". These are two different types of zeros each with different rules. One is a quantity; the other is an ordinal value.

What Difference Does It Make?

Why is the correct view of zero important? In our day to day lives, it makes little or no difference. If fact, it is remarkable how little difference it makes to our everyday existence. Although we may talk about dividing 2 apples between 0 people, for example, it is not something we ever need to do. Similarly, no one has ever made or lost money because of whether 0 divided by 0 is undefined or not.

This is not just a philosophical issue however. It makes a huge difference in theoretical physics. Interpretations of mathematics from the Big Bang, Singularities, String Theory and the nature of dimensions are impacted. For example, whether the laws of physics break down in a singularity or not depends solely on whether zero is treated as the presence or the absence of a number.

Not only are our usual conventions for zero illogical, leading to results such as that division by zero is "undefined" or "infinity", they do not match the physical universe we live in. This following of inherited mathematical conventions relating to zero has, in turn, coloured the views of generations of physicists. Even luminaries like Einstein were baffled by the nonsensical results their equations sometimes produced when they applied the usual interpretations of the day.

As we shall see our interpretation of the humble zero decides things such as:

- Whether the laws of physics break down in singularities.
- Whether worm-holes exist or not.
- Whether the Big Bang exploded out of nothing.

Many of the big issues in physics are based around zero and especially division by zero. In some of these cases we are not even aware that we are involving division by zero, yet we do. In these cases we apply the rules of a 1-dimensional value to a 0-

dimensional value. This alone causes much of the confusion that surrounds zero.

We commonly apply the rules of ordinal "real" numbers to zero. This is correct when using zero as an ordinal value, but not for physical quantities or any other non-ordinal use of zero. When using the "nothing" view of zero and treating it as the absence of a number, all of these issues literally disappear. They disappear because they are merely the result of a mathematical error.

Useful Definitions

Before continuing, let's define a few terms we will use later in this book.

Placeholder, Ordinal or Quantity

As discussed, there are several different uses of zero, such as a placeholder in our numbering system, as an ordinal number and as a physical quantity (also called a cardinal number).

Unless otherwise specified, the term "zero" used on its own will denote the use of zero as a quantity, rather than as a placeholder or ordinal value.

"Nothing" and "Number Zero"

In the following chapters we will use "Nothing" to denote the absence of any value and "Number Zero" to denote the use of zero as an ordinal value.

Using set theory we can define these as:

"Number Zero": { 0 } Zero is treated as the set of a single integer half-way between -1 and 1.

"Nothing": { } The empty set or the absence of any number.

"Number Zero", as it is commonly used as an integer between -1 and 1, is not the same as "Nothing". It is our convention to treat it as a number, much the same as 1, 2 or any integer we could place on a number line. We commonly use it as the presence of a value, rather than the absence of a value both for ordinal numbers and quantities.

"Nothing" is the absence of anything. The lack of any number at all. A zero-dimensional value. An empty set.

For example, compare 0 time (no hours, minutes or seconds) with the hour 0:00 – 0:59 on a digital clock. 0 time has no duration. Our ordinal "hour 0" lasts 60 minutes.

Geometric Points

We will discuss points later in this book. A point is a location in space, on a line or on a plane, which has no extent. That is, it has zero width, zero height and zero depth. It is a dimensionless object having no properties except location.

"Undefined"

The results of certain mathematical operations are described by mathematicians as "undefined" or sometimes called "disallowed". 0 divided by 0 is one such operation conventionally referred to as "undefined". "Infinity plus infinity" is also "undefined".

This does not mean that "infinity + infinity = 0 / 0", just that both are "undefined" operations.

"Undefined" does not mean that the result is actually a value called "undefined". "Undefined" is mathematician-speak for "I don't know", it is not a number.

4 Limits and Scope

Many calculations have in-built limits although we do not always explicitly state them. These are values beyond which the calculation returns no useful result. Many of the examples we will look at later in this book are capable of going "out of scope" or beyond their limits, so let's take a look at this before proceeding.

The Scope of Equations

Limits can apply to everyday operations such as x / y. If x = 1 and y = 0 then this is 1 / 0. The conventional view using zero as a number is that this must be either "undefined", which is not a number, or infinity, which is also not a number. How could we get not-a-number from 2 numbers and why is this the only case in which it happens?

It is not immediately apparent, but 0 is out of scope for x / y. If we wrote 1 / pineapple, it is clear that the divisor is not a number and that the result is therefore also not a number. The reason that we do not notice that zero is out of scope is because for the other basic arithmetic operations: x + y, x - y, x × y it makes no difference whether you treat zero as a number of not.

If zero is not-a-number and not-a-number is zero then 1 × 0 is zero no matter whether you treat zero as a number or as nothing. It only makes a difference whether you treat zero as nothing or a number, or as in-scope or out-of-scope, in division.

Another example of a limit is: $x = SQRT(y)$. It is well known that any attempt to take the square root of a negative number produces an "imaginary number". This is a case where an input value is clearly outside the scope of the equation. It has gone beyond the point where it produces a reliable result. For any value of $y > 0$, we get a valid result. For any value $y < 0$ we do not. In the case where y is less than zero, we cannot find the square root, as the square root of a negative number does not exist. Implicit in the calculation is: $x = SQRT(y)$ where $y > 0$.

Many times, equations and mathematical expressions have logical boundaries. Sometimes an equation is only valid under a limited set of conditions, although these are not explicitly stated or immediately apparent.

There are also many less obvious limits however. For example, if we want to measure how full a container is we could use:

```
fullness = contents / capacity
```

Let's take a 2-litre container. When can calculate how full it is by:

```
fullness = contents / 2
```

When it's empty it is 0% full.
If we add one litre it is 50% full.
If we add another two litres is it 150% full?

No, of course not. There is an in-built limit to our equation. Any value of `contents` which is greater than 2 litres is simply out of scope. Any value of `contents` which is greater than `capacity` is beyond the ability of our equation to return a reliable result.

Correctly stated, our equation for fullness is:

```
fullness = contents / capacity
where contents > 0, capacity > 0 and contents ≤
capacity
```

What a mouthful! You see why explicitly stated limits are so often omitted. Note that I have written `contents` > 0, not `contents` ≥ 0. This is intentional. This use of zero is as a physical quantity, not an ordinal. In this case zero is not a number and plays no part in the result. We will examine this in greater detail later.

Some well-known examples of results in physics that are commonly read for out-of-scope values include the Lorentz Factor (which is used in Relativity) and density calculations (such as in Black Holes). In the case of the Lorentz Factor values of v ≥ c are out of scope. Similarly zero volume is outside the scope of density equations. Why? Because zero volume is not a volume–it's a point. An equation that works for a 3-dimensional value does not work for a 0-dimensional one.

Infinities resulting from division by zero in Physics are generally the result of an equation that has gone out of scope. Results of zero may also come from an equation that has gone out of scope, although this is often not as obvious. The main point is that there is no information contained in the result–it is the absence of any value, not an actual number.

Scope and Rotations

In later sections we will look at "rotations", so before we do let's look at issues of scope with rotation formulas.

Any rotation through 360 degrees can be divided into 4 equal 90 degree quadrants without any quadrant overlapping any other:
  (a) From 0 degrees up to but not including 90 degrees.
  (b) From 90 degrees up to but not including 180 degrees.
  (c) From 180 degrees up to but not including 270 degrees.
  (d) From 270 degrees up to but not including 360 degrees.

Figure 4   Rotations and Quadrants

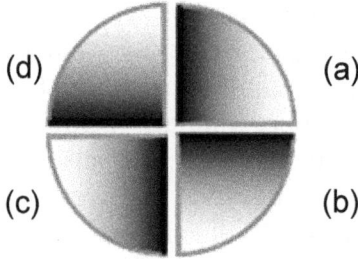

When we look at "rotation" equations we need to bear in mind that there is no overlap between one quadrant and the next. For example, the first quadrant starts from 0 degrees but does not include 90 degrees. The start of the next quadrant is out of scope of the previous one.

One way to view this is by way of a 2-dimensional object like a triangle. Let's view it from a range of angles plus or minus 90 degrees. When we view a 2-dimensional object "edge on" at 0 degrees there is nothing to see. When the shape ceases to be a triangle and becomes a line with zero width, then the result is null or "nothing".

Figure 5   Triangle View of Rotations

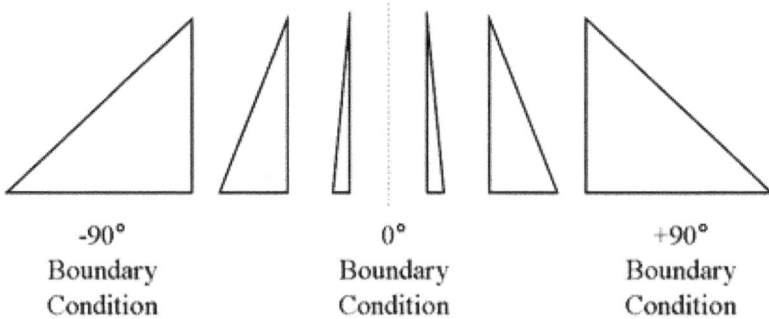

-90°	0°	+90°
Boundary	Boundary	Boundary
Condition	Condition	Condition

If the object we are examining is 2-dimensional, like the triangle then we encounter boundary conditions at -90 degrees, 0 degrees and +90 degrees. Through a rotation of 90 degrees from -90 degrees to 0 degrees, we are now viewing it "edge-on". Regardless of how large the triangle may be, or what shape, edge-on it has zero dimensions. We can tell nothing about it at all.

Later we will look at the examples of boundary conditions issues in such cases as in the slope of a line, in singularities and in a rotation from Special Relativity.

5 Zero and Mathematics

A look at the various types of zero we use in mathematics. We'll look at the confusion that is caused by the lack of a solid conceptual view of the zeros we use, because it is here that mathematics with zero begins to unravel.

Rules For Ordinals and Quantities

One of the defining features of ordinal numbers is that we do not normally multiply or divide by them. We often add to and subtract from ordinal values, but we do not use them in division or multiplication in everyday use.

We often add quantities of time to ordinal values. For example 11:00 AM + 1:30 = 12:30. However, although we may divide distance by time in hours or minutes we would never divide 10 kilometres by the time 12:30 AM – it is an ordinal value, not a quantity.

There is little difference between the "Nothing" { } and "Number Zero" { 0 } views for addition, subtraction and multiplication except the difference in views as either the presence or absence of a value. There is a considerable difference however, for division.

If, when dividing or multiplying quantities, zero really represents "nothing at all" then there is literally nothing to operate on and the result is "null" or nothing, rather than any value. (The examples below work equally well with 2+0, 3-0, 4×0, etc, I use 1 for simplicity and consistency).

Using the conventional "Number Zero" view:

```
1 + 0 = 1
1 - 0 = 1
1 × 0 = 0
1 / 0 = Infinity, impossible or undefined
```

Viewing zero as "nothing", the empty set or the absence of any value, these instead become:

```
1 + { } = 1
1 - { } = 1
1 × { } = { }
1 / { } = { }
```

Because $0 = \{ \}$, this is the same as:

```
1 + 0 = 1
1 - 0 = 1
1 × 0 = 0
1 / 0 = 0
```

Note the variation in the results between the views above: division by zero. This is the only one of the basic arithmetic operations that is controversial.

Using the view that zero is not a number and it is the absence of any value, then $1 / 0 = 0$. Not infinity, impossible or undefined. There is no number to divide by, so the result is not a number. Taking the "Nothing" view, we interpret zero as an absence of a number and we also interpret the absence of a number as zero, meaning that $1 \times 0 = 0$ and $1 / 0 = 0$.

The table below shows a side-by-side comparison of the results of basic operations using zero as a number { 0 } and as the absence of a number { }.

Operation	0 as "Number Zero"	0 as "Nothing"
Addition	$1 + 0 = 1$	$1 + 0 = 1$
Subtraction	$1 - 0 = 1$	$1 - 0 = 1$
Multiplication	$1 \times 0 = 0$	$1 \times 0 = 0$
Division	*Infinity / Undefined / Impossible*	$1 / 0 = 0$

The "Nothing" view then gives two differences from the conventional view:

(a) Results may be absent or "incomplete" reflecting the absence of a value, and

(b) The result of division by zero is zero or "nothing".

Interestingly, the old axiom that you "should not divide by zero" still holds true, not because it arbitrarily avoids infinities but because there is nothing to divide by and it is the right thing to do.

The Infinities

No discussion of zero is complete without discussing its distant cousins–infinity and infinitesimal. Infinity is the term we commonly use for uncountable things or immeasurably large quantities. Infinitesimal is the inverse–infinitely small but greater than zero.

If zero can be considered to be the point before the start of the number line, then infinity can be considered to be the point after the end of the number line (although it doesn't actually exist as the number line doesn't have an end).

Note that infinity is not a number–it is a concept. Similarly, infinitesimal is not a number, it is just a term to describe something arbitrarily small.

There are at least three distinct types of infinities in common use, although they are commonly treated indiscriminately. To differentiate between these let's call them "Type 1", "Type 2" and "Type 3".

Infinity Type 1

Our "Type 1" infinity is the number of parts we can divide something into. It is simply an unbounded quantity with any value other than zero. We can define "infinitesimal" to be as arbitrarily large or as small as required, as long as it is not equal to zero. This type of infinity is the same whether zero is viewed as { 0 } or { }. This type of infinity is sometimes called a "potential" infinity: it can always be made larger but is, for any particular value, actually a finite number.

Example:

```
1 / infinitesimal = infinity
```

Is a "Type 1" infinity actual infinite? Let's take the example:

```
x = 1 / y
```

We see that as we make "y" smaller and smaller (approaching 0), "x" becomes larger and larger (approaching infinity). This is the reason we treat this type of result as an infinity–because there is no limit to its size. We can always make "x" larger by making "y" smaller. Note, however, that as soon as we have a fixed value for "y", then "x" is actually a finite number, not infinity. This is what the Greek philosopher and mathematician Aristotle termed a "potential infinity" rather than an "actual infinity".

Infinity Type 2

Our "Type 2" infinity is generated by division by zero. It only occurs when zero is viewed as a number. It is generated by division by zero and only by division by zero. The logic behind it is that if zero is a number then: Subtract zero from one, if the result is not zero, repeat. If zero is viewed as the absence of any number, like the empty set { }, then there is no division and no "Type 2" infinity.

Example:

```
1 / 0 = infinity
```

Infinity Type 3

The "Type 3" infinity can be best described as "the set of never-ending things". The number of integers, never-ending distances and infinite loops also fall into this category. This is much the same whether zero is viewed as a number or as the absence of a number.

Example:
```
The number of odd integers = infinity
```

The Same but Different

The various types of infinities have many things in common and some subtle differences. For example, here we differentiate "Type 1" and "Type 2", which are the direct result of division, from "Type 3" which can be described as a simple endless loop, limited only by the available time and space.

Also, contrast a "Type 1" infinity with a "Type 2" infinity. One might say that a "Type 2" infinity is infinitely larger than a "Type 1" infinity, as zero is infinitely smaller than infinitesimal.

In calculations, "Type 1" and "Type 3" infinities can generally be replaced with the words "any number". "Type 2" infinities do not follow this rule–they are the result of a division by zero.

Although our system of "Type 1", "Type 2" and "Type 3" infinities overlaps with the "countable infinities" and "uncountable infinities" of Georg Cantor and the "potential infinities" and "actual infinities" of Aristotle, we shall stick to our own definitions here to avoid any confusion.

Division By Zero In Mathematics

You will often find quantities such as time or distance, occurring in equations, but you never need to divide a quantity by an ordinal value. For instance when dividing distance by time we may see: distance / 1 hour, but never distance / 1:00 AM or distance / 3:45 PM. Why not? Because ordinal numbers are a human invention, not natural quantities.

We never need to divide by an ordinal number zero because we invented the ordinal number zero to suit ourselves. We used it to plug a gap between -1 and 1 in our range of integers, or between 23 and 1 in our system of recording time, and then we insisted that "nothing" must follow the same rules. It was here that mathematics went wrong.

From this point in time we became stuck with the ideas that any number divided by 0 must be either: "infinity", "impossible" or "undefined". The real answer is that division by zero (nothing or the absence of any number) results in zero (nothing or the absence of any number).

Consider the simple case of one divided by zero (1 / 0):

The conventional view of division by zero is that it is an operation for which you cannot find an answer, so it is "disallowed".

One reason that 1 / 0 is considered "disallowed" or "undefined" is it that it has no multiplicative inverse. For example, if we assume that zero is a number and that 1 / 0 = Infinity, this implies that Infinity $\times$ 0 = 1. This requires that 0 + 0 is greater than 0.

This requirement for a multiplicative inverse is because for all real numbers division and multiplication are related–one is the inverse of the other.

```
12 / 3 = 4
```

and

```
3 × 4 = 12
```

However, if zero is the absence of a number:

```
12 / 0 = 0
```

but this does not mean that:

```
0 × 0 = 12
```

The assumption that division by zero doesn't work only occurs if you insist that zero is a number and that it must have a multiplicative inverse, other than itself.

The conventional view of zero is that it is a real number. This is what leads to conclusions such as that 1 / 0 is "undefined" or "impossible". There is an alternative view, that zero is the absence of a number, not an actual number. In this case we treat zero as the empty set { }, nothing or the absence of any value at all.

If zero is a number then 1 / 0 gives:

A number divided by another number equals "not-a-number".

This is an exception in mathematics. Any other number will produce a number as a result. How is it possible to produce "Not a Number" when dividing one number by another? Why would it only happen with zero? When treating zero as nothing, the rules are clear and simple: zero is not a number, therefore division by zero does not return a number.

The alternative, that zero is not a number means 1 / 0 becomes:

A number divided by not-a-number equals not-a-number.

This removes the exception from mathematics.

This is the only difference that arises among the operations (1+0, 1-0, 1×0 and 1/0) when viewing zero as the absence of a number, rather than as a the presence of a number. 1+0 still equals 1, 1-0 still equals 1, 1×0 still equals 0.

Using the "nothing" view of zero does not change the outcome of addition, subtraction or multiplication, only division. The only difference is that the "controversial" issue of division by zero is resolved.

Is Division By Zero Impossible?

The reason we concluded that division by zero is impossible or undefined or infinity is because we treat zero as a number. If we treat zero as the absence of a number and apply a suitable set of rules for dealing with the absence of a number then we will come to a completely different conclusion, namely that $1 / 0 = 0$.

We do not think that division by 1 is impossible or undefined or that subtraction of zero is impossible or undefined so what is so special about division by zero?

As we shall see in the Physics section later, Nature is confronted with division by zero regularly, so it does not have the luxury of "undefined". It is possible to tell a student that division by zero is "disallowed", but you cannot dictate to the mathematics of the Universe.

Of the conventional interpretations about division by zero "impossible" is the probably the closest to the truth in that if there is no number then there is no number as a result. This absence of a number is zero, so the actual outcome of $1 / 0 = 0$. This is not strictly impossible then, it is just impossible to get a number as the result–the result will always be the absence of a number.

0 Divided By 0

If there is a single operation that should flash red warning lights to indicate there is a problem with our conventions for zero it is the seemingly trivial 0 / 0.

- Ask a child what nothing divided by nothing is and they'll tell you it is nothing.

- Ask an ancient Babylonian mathematician what nothing divided by nothing is and he would have told you it is nothing.

- Ask most modern mathematicians what 0 / 0 is and they'll tell you it is "undefined".

- Ask some other mathematicians what 0 / 0 is and they'll tell you it is infinity.

- Any number divided by itself is 1.

So, is 0 / 0 equal to nothing, 1, "undefined", infinity?

The theory that 0 / 0 is "undefined" arises thus: If 0 / 0 = 0, what value, when multiplied by 0, equals 0?

Modern mathematicians seek to find a single result that provides a multiplication that is the inverse of the division. For example, the inverse of 1 / 2 is 2 × 1. The result is generally called "undefined" as it is held that there is no single result that is the multiplicative inverse for 0 / 0.

Unfortunately, 1 × 0 = 0 and 2 × 0 = 0 and any other number multiplied by 0 equals 0. The conventional view is that because there are an infinite number of multiplicative inverses, then 0 / 0 must be "undefined".

Remember that "undefined" does not mean that the result of 0 / 0 is actually a value called "undefined". "Undefined" is just mathematician-speak for "I don't know". Note once again that the problem here involves zero. We are not arguing whether 2 / 2 is undefined nor 100 / 100.

The answer that 0 / 0 is "infinity" comes from the view that we can subtract 0 from 0 an infinite number of times, although this is a minority opinion.

The conventional view of "Number Zero" is simply flawed and falls apart when you apply it. Once again this is a problem caused by our conventional treatment of zero as a number, rather than as the absence of a number. This leads to strange and confusing results.

Using the "Nothing" view of zero, 0 / 0 = 0. Always.

As we shall see in later sections, the mathematics of Physics does not have the luxury of "undefined" or "infinity". The Universe doesn't suddenly get confused when it encounters a zero and it doesn't produce unpredictable results.

We will see the result of this absence of any value by looking at those cases in the mathematics of Physics where we find 0 / 0. We find this in Newtonian Physics, Special Relativity and in Planck's Constant. We'll examine these in more detail in the Physics section.

Does 1 = 2?

One of the classic games played with "number zero" is to prove that $1 = 2$ (or $2 = 5$ or anything else).

The game is played like this:

(1) If we start with the following assumptions:
```
0 × 1 = 0
0 × 2 = 0
```

(2) And then assume that the following must be true:
```
0 × 1 = 0 × 2
```

(3) If we divide both sides by zero we get:
```
0/0 × 1 = 0/0 × 2
```

(4) Simplified, this gives us:
```
1 = 2
```

So where did we go wrong?

In step (2) there are no values on either the left or right sides of the equations.

```
0 × 1 = 0 × 2
```

is exactly equivalent to:
```
{ } = { }
```

There are no values on either side.

It's the same again in step (3).

$$\{ \ \} = \{ \ \}$$

There are no numbers on either side of the equation. Note that as soon as we attempt any division by zero the result is always the absence of a number.

Step (4) is a completely different situation. This step does not just simplify the results, it removes the empty sets. In this case we are left with numbers, although they are arbitrary and meaningless and are completely unrelated to the results we had from step (3).

Note that throughout steps (1), (2) and (3), the numbers (1 and 2) played no part in the results. If zero is not a number there is no result.

Reciprocals

All real numbers are generally considered to have a reciprocal or "multiplicative inverse". That is, if we can divide by a number then we should be able to multiply by it. The product of the reciprocals of any two numbers is 1.

The multiplicative inverse of a/b is b/a. For example:

```
a / b = c
```

Therefore:

```
b × c = a
```

We expect this of numbers. For example the reciprocal of $1/2$ is $2/1$. Both results are numbers and their product $(2 \times 1/2)$ is 1.

However, when using the conventional view of Number Zero:

```
0 / 1 = 0   and   1 / 0 = ?
```

$0/1$ is a number (zero) and $1/0$ is not (it is undefined, impossible, infinity, depending on your opinion).

When treating zero as "Nothing" or the empty set:

```
0 / 1 = 0   and   1 / 0 = 0
```

Or:

```
{ } / { 1 } = { }   and   { 1 } / { } = { }
```

There are missing inputs, and the absence of any output. Neither of the results is a number and $a/b = b/a$. Once again, the result is that there is no multiplicative inverse for zero, other than itself.

The Riemann Sphere

In mathematics, a Riemann sphere can be thought of as a complex number plane wrapped around a sphere, with one additional point at infinity. It is named after 19th century mathematician Georg Friedrich Bernhard Riemann (1826–1866).

Figure 6   The Riemann Sphere

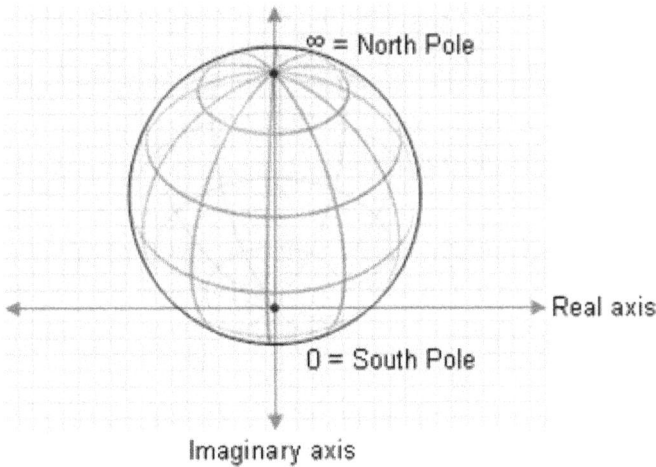

The Riemann sphere is generally described as a way of extending the plane of complex numbers, in a way that makes expressions such as 1 / 0 = infinity better-behaved and useful. It is often regarded as having "tamed" zero and infinity. This is actually far from the truth.

One of the features of the Riemann sphere is that numbers on opposite sides of the sphere are reciprocals. That is, if we take 2 for example, on the opposite side of the sphere we will find 1 / 2; the reciprocal of 2. It is a commonly-held view that because we can place zero at one pole of the sphere and infinity at the other then they must be reciprocals of each other.

There are several problems with using the Riemann sphere in conjunction with zero and infinity, including:

- The complex number plane is infinite, whereas a sphere has finite boundaries.
- Neither zero nor infinity are numbers and they do not actually exist on the number plane, whether it is wrapped around a sphere or not.

You may hear that because we can place a zero point on one side of the Riemann sphere and an infinity point on the opposite side that this "proves" that 0 × infinity = 1. Once again, however, we are dealing with a man-made tool, designed to simplify complex concepts, not the laws of nature.

You may read comments about the Riemann sphere such as "infinity becomes a point like any other number". The truth of the matter is that real numbers have extent and are not points.

On the number line, zero may be viewed as being the point before the start of real numbers and infinity the point beyond real numbers. In both of these cases, zero and infinity, these points are out of scope for reciprocals. Remember that infinity is a concept, not a number. We may as well locate a point on the sphere and call it "pineapple"–the result is equivalent–our point is not a number.

However appealing and useful the Riemann sphere is in certain circumstances, it is an abstract representation, not a mathematical law. If 1 / 0 = infinity and 0 × infinity = 1 then it follows that 0 + 0 > 0; which is an obvious absurdity.

Using set theory, we can say that:

{ } × 1 = { } and { } × 1000 = { }

An empty set never interacts with numbers no matter how many times you try it, even an infinite number of times. This leads us to the conclusion that infinity times zero equals zero: $\infty \times 0 = 0$. That is, zero has no reciprocal number other than itself.

Note that once again, the controversy does not involve 1, 2 or any other non-zero number. There can be no doubting that 1 / 2 is the reciprocal of 2 / 1 or that 1 / 100 is the reciprocal of 100 / 1. The problem once again involves the use of zero as a number.

So although the Riemann sphere may be useful in many contexts, it is not an actual representation of either zero or infinity. Like the number line, it is a much-abused tool and leads us to believe that zero-sized points off the number line are actually numbers.

Zero Raised to the 0th Power

For an example of the vagaries and effect of opinion in mathematics let's look at the seemingly mundane topic of zero raised to the 0th power (0^0, 0^0).

Many have attempted to prove that $0^0 = 1$, generally on the basis that every non-zero number raised to the 0th power equals 1, although opinion is divided.

Review any history on this topic and you will see how much of popular mathematics is "indeterminate", "controversy" and "consensus". Below is a brief sample of views from Wikipedia:

> *"Some argue that the best value for 0^0 depends on context, and hence that defining it once and for all is problematic. According to Benson (1999), "The choice whether to define 0^0 is based on convenience, not on correctness. "*

> *Others argue that 0^0 is 1. According to ... Knuth (1992), it "has to be 1", although he goes on to say that "Cauchy had good reason to consider 0^0 as an undefined limiting form" and that "in this much stronger sense, the value of 0^0 is less defined than, say, the value of 0+0". "*
>
> http://en.wikipedia.org/wiki/Exponentiation#Zero_to_the_zero_power

This is another case of "undefined" in mathematics because of lack of clear understanding of what zero is. Opinions and convenience are fine for certain subjects but personally I prefer my mathematics to be based on, well... mathematics.

$0^0 = 0$: An Ordinal and a Quantity

This expression contains both the ordinal and quantity uses of zero. In the example A^0 (A^0) "A" raised to the power of zero, "A" is a quantity and "0" is an ordinal value–a human-invented convention for representing a power to which we wish to raise a number.

Many people fail to distinguish between the ordinal and quantity uses of zero in this expression. This is another case when the lack of a clear understanding of zero can be misleading. Let's examine this below using sets:

We can represent 1^0 as: $\{ 1 \}^{\{ 0 \}}$
We can represent 0^0 as: $\{ \ \}^{\{ 0 \}}$

As an ordinal number, as in "to the zeroth power", we use zero as the set of a single integer halfway between -1 and 1: $\{ 0 \}$.

As a quantity zero is an empty set $\{ \ \}$ we wish to raise to this 0th power.

Each of these uses of zero has different rules.

When we incorrectly view the quantity as "number zero" $\{ 0 \}$ we assume there is a number to work with and confusion arises.

When using the "nothing" $\{ \ \}$ view of the quantity we assume there is no number. "Nothing" or a null quantity cannot be raised to any power as there is nothing to raise, so the result is 0. This is a result based on mathematics, not convenience or consensus.

Note that the result for zero versus non-zero quantities also reflects the number of dimensions in the quantity. A^0 when A is any non-zero number, is agreed to equal 1. This result of 1 is the same as the number of dimensions in our quantity "A". Taking the "nothing" view, when A=0, nothing cannot be raised to any power, so the result is null or zero. Zero is also the number of dimensions in our quantity "A".

6 Zero and Geometry

In this section we take a look at some common misconceptions about points, lines, planes and solid objects. These misconceptions are not only found in geometry but also in physics and generally involve the misuse of zero-sized objects to construct non-zero sized objects.

Points, Lines, Planes and Cubes

A quick review of the Euclidean geometry we learnt at school:

- A point has no width, height or depth
- A line has width but not height or depth
- A plane has width and height but not depth
- A 3-dimensional object, such as a cube, has width, height and depth

This leads us to:

- A line cannot be constructed from points.
- A plane cannot be constructed from points or lines.
- Solid objects cannot be constructed from points, lines or planes.

In the following sections we will examine this in greater detail.

The Problem With Points

Mathematical points encompass no area, no volume, no space and no time. They are mathematical abstractions that do not exist in the real Universe. A volume of 0 x 0 x 0 is mathematically equivalent to a point (also 0 x 0 x 0) or a square with an area of zero (0 x 0 x 0). In each case the value represented is the absence of anything: no area and no volume.

The sole purpose for points is to locate objects: to specify the co-ordinates at which we will find the object. The actual object is never itself a point. It must have some attributes: size, duration, mass, etc. A dimensionless point can only ever describe the dimensionless object "nothing".

Points in a Line

If we think of zero as just another number, then it is easy to overlook the fact that a line has length, but no height or depth and that a point has no dimensions at all. Any attempt to divide a line into zero-sized points is a case of division by zero as a point has zero width.

Constructing a Line Versus Locating Points on a Line

It is a common misconception that you can create a line using only an infinite number of points.

Aah, I hear you say, but I can place a point 1 / 2 of the way along the line, then 1 / 3, 1 / 4, 1 / 5, 1 / 6 and so on and thus place an infinite number of points on the line! But this is not constructing a line from zero-sized points, it is placing points a non-zero distance apart and actually constructing nothing–there is still no line, only zero-sized points that lack any length at all. No matter how many points you place along the line, the sum of their widths will always be zero.

It is possible to LOCATE an infinite number of points on a line, a short distance apart (a "Type 1" Infinity), but that is a different matter to creating a line solely composed of points. Just because a line can be defined using points does not mean it can be created from them.

In the example below we can locate an unlimited number of points along the line. We can make the points as close together as we wish (a "Type 1" infinity). Note that we do not actually construct a line though. We merely locate zero-sized points along its span. The points themselves contribute no length to the line.

Figure 7   Locating Points On a Line

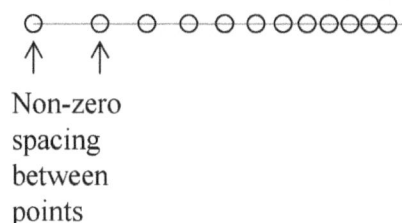

Non-zero
spacing
between
points

Constructing a Line From an Infinite Number of Points

Let's take an infinite supply of zero-sized points. Starting at the left, place a point, then move 0 millimetres to the right and repeat. Keep stacking up zero-sized points until you get to 1 cm.

Try building a line using only zero-sized points and you soon realise it cannot be done. No matter how many points you add, even an infinite number, you never move from 0.

If we try to place an infinite number of points zero distance apart we get the result shown below. All the points are located at the same place on the line. No matter how many times you do this, you will never even begin to construct a line. The only way to construct a line is from shorter segments of a line, not from points.

Figure 8   An Infinite Number of Points Zero Distance Apart

An infinite number
of points spaced 0
distance apart

We could repeat this test in an attempt to use lines to construct a plane or planes to construct a cube. The outcome would be the same–it would not work. It is simply not possible to create anything from zero-sized quantities, no matter how many you use.

Division By Zero In Geometry

Below we will look at simple divisions in geometry. Let's look what happens when we divide a line into 2 or ten parts.

Division by 2

First let's see what happens when we divide a line into two parts.

Before Division

_____

After Division

_____ _____

Note that this operation succeeds—we have divided our line into two parts.

Division by 10

Similarly we can divide a line into ten parts.

Before Division

_____

After Division

— — — — — — — — — —

Note that this operation also succeeds—we have divided our line into ten parts.

Division by 0

Now let us examine the same operation, except dividing by zero rather than two or ten. Take a line and remove a zero-sized piece (a section that is 1/0 the length of the line):

Before Division

_____

After Division

_____

Note that this operation fails. We have not divided our line into zero parts–we have not divided our line at all. Repeat this again and again. No matter how many times we attempt to divide the line by zero, nothing happens. This is what early Indian Mathematicians described when they said that "a number remains unchanged when divided by zero". No division actually takes place.

It is exactly the same as if there was no number and we had never even attempted the division. Compare that to dividing a line in two or ten (or into any other number of pieces). When dividing by any non-zero number, the line is changed: division actually takes place. When dividing by zero, the line is unchanged: no division takes place.

Dividing a Line Into Points

Let's look at the claim that it is possible to divide a line into an infinite number of points. Note that this use of infinity is subtly different from other types of infinity in that it is produced by division by zero, rather than division by infinitesimal. We shall term this type of infinity a "Type 2" infinity. Note that this type of infinity is only produced by division by zero and only when zero is treated as a number. It is never produced when zero is treated as nothing or the empty set.

Many mathematicians argue that because we can repeat this division ad infinitum and still have a remainder to divide by that the result is infinity. The actual result is failure to divide and the answer is the absence of a number. This leads us to the conclusion that $1 / 0 = 0$ (no number to divide by means there is no result) and infinity times zero equals zero: $\infty \times 0 = 0$. Put another way: $\{ \} \times N = \{ \}$, an empty set never interacts with numbers no matter how many times you try it, even an infinite number of times.

This shows some of the mythology surrounding zero and division by zero. As we have seen, you can locate an infinite number of points on a line but you cannot construct a line from points. Similarly, any attempt to divide a line into zero-sized points fails.

This also demonstrates clearly that zero does not follow the conventions of numbers. Zero refuses to play by the rules of numbers. Nature treats zero as the absence of a number and the absence of a number as zero. It treats zero as the absence of anything, not as the presence of something.

The Scope of Slope

Rotations occur in a variety of forms. In some cases it is not immediately apparent that the formula is a rotation. Take the formula for "slope" for example. This formula is for finding the deviation of a line from either the horizontal or the vertical.

To find the slope from the horizontal, we use the following formula:

```
m = (y1 - y2) / (x1 - x2)
```

Figure 9 Slope

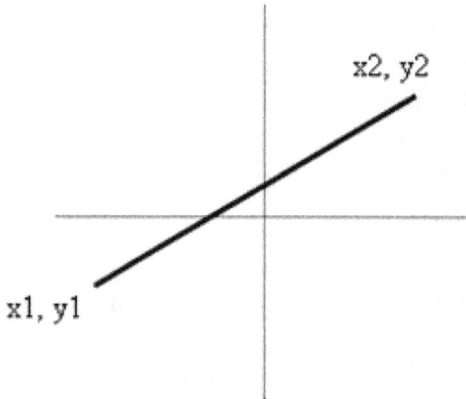

This formula is, in fact, a type of rotation. The scope of this formula is from 0 to less than 90 degrees. Note once again the ordinal use of zero here; 0 degrees occupies the first degree of our quadrant. 90 degrees is the start of the next quadrant, not the end of the first. For this reason we can measure the horizontal slope, but we cannot simultaneously measure the vertical slope.

Let's look at the case where we have a vertical line and x1 = x2 and the result is (y1 − y2) / 0. You may hear or read that "the slope of a vertical line is undefined because division by zero is not allowed", but this is not the full story. This division by zero comes about for a reason–the equation has gone beyond its useful scope and is no longer returning valid results. To measure the vertical slope we need to change the formula to:

```
m = (x1 - x2) / (y1 - y2)
```

This is the same as turning our formula around by 90 degrees.

7 Zero and Physics

We examine some of the laws of Physics and the actual world we live in and compare the usual conventions for "Number Zero" with those for "Nothing". We will look at the various ways that zero interacts with the mathematics of physics and how it is sometimes misunderstood.

Mathematics Versus Physics

As we have seen, some of the most useful mathematical tools for managing zero, such as set theory and Cartesian coordinates are fairly recent developments. Additionally, some bad or ambiguous concepts such as "infinite points in a line", "number zero", "undefined" and the zero/infinity points on the Riemann Sphere have muddied the waters.

Most of our conventions for mathematics were developed prior to the development of modern physics. In fact much of the basis of the rules we use are matters of philosophy and opinion rather than being mathematically and rigorously tied to the physics of the universe.

In effect generations of physicists have taken the mathematical and philosophical rules of their time and applied them to physics without ever asking themselves if the rules were correct. They took it for granted that the Universe operated on the rules they were taught.

Physics is peppered with problems involving zero and infinity. Some of the problems are obvious and some are more subtle. For example, the "division by zero" infinity in the Lorentz Factor used in Special Relativity at v=c is quite obvious, however the division by zero involved when modelling an electron as a point particle is less so.

These problems with zero fall into a number of categories, including:

- Treating the result of division by zero as a number.
- Not realising an equation has gone out of scope.
- The use of zero-dimensional points to model actual objects.

At the root of each of these problems is the treatment of zero as a number, rather than the absence of a number.

The examples in this section are based on the mathematics of physics rather than on the self-referential system of conventional mathematics. As we shall see the mathematics of physics is demonstrably at odds with our conventional rules of mathematics where zero is concerned.

Infinities In Physics

Next in this section we will look at some cases of division by zero involving physical properties and the mathematics of physics. Many of these cases involve quantities that may be very small, approaching zero or actually zero. This raises the question of whether these quantities may ever actually be zero or not.

For example, does space have a minimum scale, a distance that limits the resolution of one piece of space and the next? In almost every case we see that both "Type 1" infinities and "Type 2" infinities may be involved. It is useful to revisit our various types of infinities and where they fit in.

If you read a book or research paper covering a topic in physics, you may well come upon the word "infinity", but what type of infinity is it? All too often it is left to the reader to form their own opinion.

As we have outlined, there are many types of "infinities". Let's examine some uses in physics.

"Type 1" Infinities (Division By Infinitesimal)

Example – An infinite number of field lines in a magnetic field.

As the field lines cannot have a zero size, then they must have a non-zero size and this type of result is actually a finite number.

"Type 2" Infinities (Division By Zero)

Example – Division out of scope.

This includes cases such as in the Lorentz Factor when $v = c$. We will take a more detailed look at this type of "1 divided by zero" infinity later in this section. It also includes cases of "infinite density" in Black Holes when assuming matter is squeezed into zero volume. This is another "division by zero" infinity that we will explore shortly.

"Type 3" Infinities (Never-ending Quantities)

Example – Unlimited distances.

We find this in cases such as in gravity or Coulomb's Law. These describe forces with infinite range. It is a convention in some areas of physics to assume an object is infinitely distant. Note that this is a different type of infinity from "Type 1" or "Type 2".

Note also that if the universe has a closed curvature then distances may be large, but not infinite, although this is the subject for some other book...

Removing the Spurious Infinities

To rid physics of human-introduced infinities, we must first rid it of errors involving zero. Why must we do this? Because Nature treats zero as the absence of a number, not as a number. It treats any physical quantity as absent, rather than present, if it is zero.

Assuming that no physical quantity of zero is present removes the spurious "Type 2" infinities. This shows us any remaining "Type 1" or "Type 3" infinites. "Type 1" infinities may well be finite numbers. A number of "Type 3" infinities may be left. These are the ones that next need attention.

To rid a theory of superfluous infinities this process may be used:

1) Catalogue and categorise the various infinities as "Type 1", "Type 2" or "Type 3".

2) Reformulate the theory, setting limits at zero for all quantities to remove the "Type 2" infinities. Review any remaining "Type 1" infinities. Are any of them actually finite?

3) Examine the reformulated theory, especially any remaining "Type 3" infinities.

Division By Zero In Physics

Although our usual convention in mathematics is to avoid division by zero, the mathematics of physics does not always have this luxury. There are cases when certain physical properties can be zero and some defined outcome must result.

Later in this section we will see that using the usual mathematical conventions we have inherited, Planck's Constant is variable and the speed of light is sometimes undefined. That is not to say that they *actually are* variable or undefined, merely that this is the interpretation we have when using the standard conventions of mathematics with zero.

What's the problem? The problem is that mathematics with zero works slightly differently to our conventions for real numbers. When treating zero as the absence of a number, we find that Planck's Constant remains a constant and that e can equal mc^2 without the speed of light sometimes being "undefined" or "impossible".

The view of zero quantities as "nothing" also highlights the use of equations which have passed the point of returning useful results, that is when they have gone out of scope.

Planck's Constant

Let's look at zero and a physical constant. Planck's Constant is a physical constant that relates the energy of photons to their frequency. Its value is 6.626 068 96 x 10^{-34} J s.

It is generally used as: E=h×v where E is energy, h is Planck's constant, v is frequency. For example, the energy of a photon is its frequency times a constant. What happens when both frequency and energy are zero? What is h=E/v when v=0 and E=0?

Calculation	Result when zero is { 0 } for quantities	Result when zero is { } for quantities
h=E/v when v=0 and E=0.	h = { 0 } Planck's Constant = Zero	h = { } Planck's Constant plays no part in the results
E=h×v E is energy, v is frequency, h is a constant.		

By viewing zero as Number Zero { 0 } and forcing a result, the answer is zero, meaning that Planck's Constant is not a constant– it's a variable. Viewing zero as null says the equation is null or "incomplete" and the result is similarly null or incomplete–there is no value of E/v to assign to h. Using the "Nothing" view provides a workable alternative and a constant remains a constant.

Completeness and Incompleteness

The conventions in mathematics generally assume that there is an answer to every operation, even division and multiplication. This is not the reality. If there is no input, there is no output. We often don't cater for cases where a value is entirely missing and hence the result is null. In these instances we can use the "nothing" interpretation of zero: { } or a complete absence of any number.

Let's take a look at our assumptions about the results of arithmetic operations, especially where one or more operands may be an empty set { } rather than as the set of a single value { 0 }.

The Presence or Absence of Value

Taking the { 0 } view, zero plays no part in addition, subtraction or multiplication, only in division. In the case of addition and subtraction, the result is as if zero did not exist at all. In the case of multiplication, zero nullifies the result.

Taking the { } view, zero takes no part in division either. In the null view, zero is ineffective for addition and subtraction and for multiplication or division it returns nothing. Stated another way: zero quantities take no part at all in mathematics.

Any quantity of zero is null and is the absence of any value. It will play no part whatever in addition, subtraction, multiplication or division.

Assignment of Results

We generally assume that an expression such as a/b returns a result. When we examine the mathematics of physics however, we see that this is not always the case.

Let's look at Newton's Second Law of Motion
```
force = mass × acceleration
```

which can be re-arranged as:
```
mass = force / acceleration
```

This is uncontroversial most of the time, but what happens when there is no force acting on the mass? For example, when I'm driving my car and I lift off the accelerator and maintain a constant speed?

This is the case where there is no force acting on the mass. If the mass of the object has an actual value what does Nature do when there is no acceleration? Let us assume that the Universe does not have the luxury of being "undefined".

In this case the result is:
```
mass = 0 / 0
```

or an empty set divided by another empty set:
```
mass = { } / { }
```

Cases such as this are sometimes used as examples of why, using the conventional view of zero as a number, 0 / 0 is "undefined". However, there is a much simpler explanation: No assignment takes place because zero is not a number as far as "natural" mathematics is concerned. We also see this behaviour with equations involving constants such as Planck's Constant and the speed of light.

As another example, let's look at e=mc^2:

```
energy = mass × c × c
```

which can be rearranged as:

```
c = SQRT(energy / mass)
```

What happens when there is no energy and no mass? Does the speed of light change when energy and mass are zero? Unlikely. If 0 / 0 is "undefined" then does "undefined" mean the speed of light? What, according to conventional mathematics, is the square root of "undefined"?

The only rational explanation for this is that no assignment of zero takes place in the mathematics of physics. Nature treats zero as the absence of a number: there is no number to assign and constants remain constant. Similarly, the square root of 0 is 0, meaning "no number".

As with Planck's Constant when energy = 0, the results of energy / mass exactly equals an empty set when either quantity is zero. No assignment occurs. We see this behaviour both with constants (such as c or h) and variables (such as mass or energy). This is quite unlike our conventional thinking and in computer software the inability to ever assign zero to a variable would be a disaster.

So how does Nature deal with division by zero? Nature simply ignores it. It is just the way the Universe does things and we had better get used to it, because it isn't going to change.

Density of Electric Current

Let's look at the example of current flow in an electrical circuit. We are interested in the current through a 2-dimensional cross-section of wire, so:

```
Current Density = Current / Area
```

We have a wire with a cross-section measured in sq mm and a current of 1 Ampere. We can represent the current density through the wire as Amperes per sq mm (A / sq mm).

```
1 A / 2.0  sq mm = 0.5 A / sq mm
1 A / 1.0  sq mm = 1.0 A / sq mm
1 A / 0.5  sq mm = 2.0 A / sq mm
1 A / 0.25 sq mm = 4.0 A / sq mm
1 A / 0    sq mm = ??? A / sq mm
```

If we take the wire away, so that the cross-section equals 0, what happens? Does the current density rise to infinity?

If we reduce the wire to 0 mm the current does not remain at 1 Amp, it also becomes 0 and the entire result returns nothing. This situation of "no wire, no result" is analogous to the situation inside a gravitational singularity: if there is no volume, there is no matter.

Just as reducing to zero the cross section of the wire leads inevitably to zero current, not infinite current, reducing to zero the volume under consideration in a Black Hole leads inevitably to the elimination of matter, not infinite density. A zero-sized volume simply has no room for any matter. Nature always finds a way to defeat infinite density.

Density of Mass

Let's take the real world example of density:

```
Density = Mass / Volume
```

The outcome of this division is often taken as proof that there is a region of infinite density in a Black Hole. This conclusion was reached by Roger Penrose in the 1960s. It is based on the idea that mass is squeezed and squeezed until it is squeezed into a region with zero volume.

Taking the "nothing" view of zero shows that there is no such thing as zero volume. Zero volume is the absence of any volume and is outside the scope of this equation. It also seems reasonable that mass cannot be squeezed out of existence, which it must be if it is to fit into zero volume or the absence of any space.

The actual equation should be read as:

```
Density = Mass / Volume
    where Mass > 0 and Volume > 0.
```

For example, we calculate the mass of a charged particle to include the mass-energy in its electrostatic field. If we assume that the particle is a charged spherical shell and use the conventional view using "number zero", the energy in the field is infinite if the radius is zero. The "nothing" view instead predicts that the density is nothing if the radius is zero, i.e. it is simply outside the scope of the calculation as there is no volume.

The Lorentz Factor

The Lorentz Factor was derived in the 1890s by Joseph Larmor and is named after mathematician and physicist Hendrik Lorentz. In 1905 Albert Einstein applied it in Special Relativity to calculate the degree of time dilation, length contraction and relativistic mass of a moving object. Its general form is:

```
g = 1 / SQRT(1 - v² / c²)
```

In Einstein's 1905 papers, along the way to deriving E=mc^2, he uses the Lorentz Factor in a variety of different cases. This term is as important in Special Relativity as the better-known "mc^2" part of the equations. It describes the relationship between moving bodies. v is velocity and c is the speed of light. It is the "other half" of the e=mc^2 equation we all know. It tells us how much of e=mc^2. The factor is often referred to as gamma.

A typical use is:

$$mc^2 \left\{ \frac{1}{\sqrt{1 - v^2/c^2}} - 1 \right\}$$

Relativistic Kinetic Energy

The equation is well known to require the square root of a negative number for v > c. This can be compensated for by rotating the point of view or by multiplying by i, the imaginary square root of -1. Whether any of these methods of compensating actually match the physical universe is as yet unproven.

There is another limit in the Lorentz Factor equation that is not widely reported. This limit comes from (1- v^2 / c^2) and means that it is useful up to, but not including, v=c. Values for v ≥ c (at or beyond the speed of light) are out of scope of the equation.

Lorentz Factor Infinities

The Lorentz Factor contains both "Type 1" and "Type 2" infinities:

The "Type 1" infinity is v^2 / c^2 in that we can make v as close to c as we like, in infinitesimally small increments. We can also vary v to be as infinitesimally close to 0 as we like, although it does not lead to a "Type 2" infinity.

The "Type 2" infinity occurs when v=c and
$1 / SQRT(1 - (v^2 / c^2)) = 1/0$.

The "Type 2" infinity is never generated when zero is viewed as { } because zero is treated as null or "nothing" and the division is "incomplete".

When we view zero as nothing, then in contrast to the conventional view, when v=c the result of $1 / SQRT (1 - v^2 / c^2)$ is nothing, not infinity. This has implications only exactly at velocity v=c.

The Lorentz Factor Goes Out of Scope

There is a little-known limit in the Lorentz Factor equation. This limit comes from $(1 - v^2 / c^2)$.

The Lorentz Factor is a rotation through 90 degrees, much like the formula for slope we looked at previously, although in this case the x and y axes represent time and space. It describes an object moving at less than the speed of light (Point of View A below). As discussed earlier under "Scope and Rotations", a 90 degree rotation is valid from the start of one quadrant up to, but not including, the start of the following quadrant.

In its current form the Lorentz Factor provides valid results for v = 0 up to, but not including, v = c. That is, it provides no useful results for an object travelling at the speed of light. This means that v = c is just outside the ability of the equation to produce reliable results.

The Lorentz calculation is often used to "prove" that objects travelling at the speed of light have no mass, otherwise infinite energy is required. This is incorrect however, as the factor is beyond its scope at v=c, and makes no predictions at all for bodies travelling at the speed of light. This is not to say that photons, for example, must have mass; just that the Lorentz Factor makes no prediction either way.

It's a Matter of Point of View

The factor is good from 0 degrees up to, but not including, 90 degrees (PoV A). To cover 90 degrees down to, but not including 0 degrees, would require a -90 degree shift in our point of view (PoV B). Similarly to cover 90 degrees up to, but not including 180 degrees, (PoV C) would require a +90 degree shift in our point of view.

Figure 10 Point of View

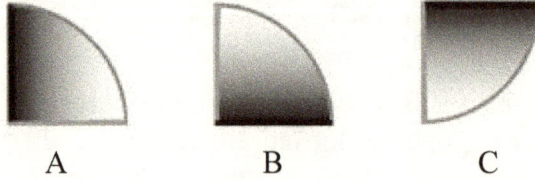

A   B   C

If 0 degrees is v=0 then v=1 is at the start of the 91st degree, not at the end of the 90th degree (note the ordinal use of zero). We can rotate our view 90 degrees to get a reliable result for v=c (Point of View B). When we rotate our view 90 degrees to see things from a photons' point of view, the result returns one (unity) for v=1 and null for v=0.

Conventional View Versus Predicted

The usual graph we see of the Lorentz Factor is shown in the figure below, with its "Type 1" infinity alongside the "Type 2" infinity at the right of the chart. On the right is the predicted version if zero is treated as Nothing (arrow).

It is a quirk of nature that the "Type 1" infinity sits alongside the "Type 2" infinity, making the two seem to be a seamless continuum when using the Number Zero view of zero.

Figure 11 Conventional View of The Lorentz Factor

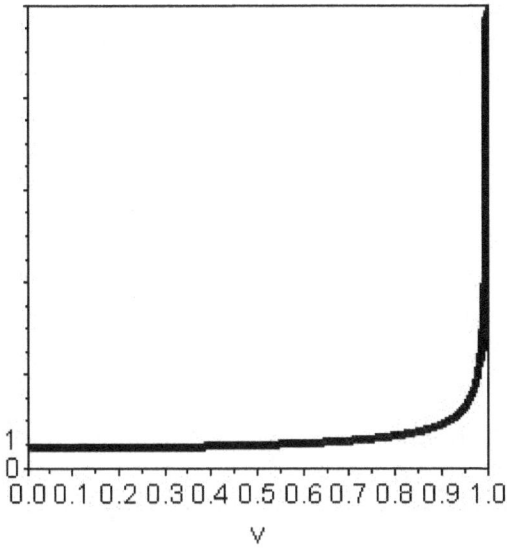

Figure 12 Predicted Lorentz Factor

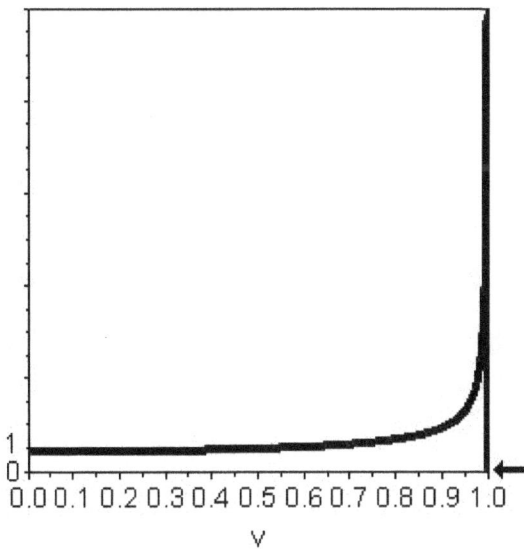

Note the only different is at v=1 (arrow).

Rotating Triangle View

As we discussed earlier in "Scope and Rotations", one way to view this result of the Lorentz Factor is to consider the v^2/c^2 ratio as a 2-dimensional shape like a triangle.

As we compare the angles "a" and "b" in the series of triangles below, the ratio a/b becomes ever larger as the triangle changes, but then the result disappears as "a" and "b" become 0. That is, at the boundary condition when the shape ceases to be a triangle and becomes a line then the result is null or "nothing", not infinity.

Figure 13 Lorentz Factor Viewed As a Triangle

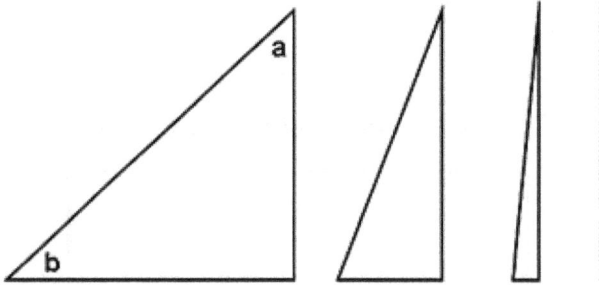

This result of zero or "nothing" is correct in that it tells us that there is no result. The reason? Because v=c is a boundary condition. If the Lorentz Factor is a rotation through 90 degrees from 0 up to, but not including, 90 degrees, then we have rotated our triangle through 90 degrees and we are now viewing it "edge-on". Regardless of how large the triangle may be, or what shape, edge-on it has zero dimensions.

Point Particles

Just as zero is the absence of any number, a point is the absence of any volume. The very use of point particles involves division by zero–space is divided into zero-sized pieces, each of which is the absence of any space. This kind of fuzziness about zero and infinity is wide-spread in modern Physics–it is the hidden "division by zero" that rarely gets a second thought.

Point particles are used to represent objects whose size, shape, and internal structure are thought to be irrelevant. The problem is that this use obscures the actual size, shape and structure and provides distorted results, such as infinities which must then be removed.

For example, modelling an electron as a point particle, which is often done, requires that it has all its energy in the absence of any space–a division by zero which is often interpreted as a "Type 2" Infinity. We are better off modelling particles on pineapples or elephants–at least pineapples and elephants actually exist.

An object may be infinitesimally small (a "Type 1" Infinity), but it cannot be zero-sized (a "Type 2" Infinity). That is, we can locate an object at a point, but we cannot construct one from a point.

Unfortunately, Physics has a long-standing use of point particles as being some sort of idealised object akin to the ancients' treatment of the planets as perfect spheres. What began as a useful tool for mathematically modelling an object of unknown size has become part of the accepted wisdom and an acceptable thing to do.

Modelling a real-world object without volume is like modelling it without time or duration, and therefore no presence in the forth dimension. If any object exists in our 4-dimensional world, it must have at least 4 dimensions; otherwise we will have to radically redefine what a dimension is.

Something Into Nothing

Nothing demonstrates the subtle interplay between zero, infinity and reality like the following from an article about time by a professional physicist I read recently (I have omitted the names to protect the guilty):

> "...in quantum mechanics all particles of matter and energy can also be described as waves. And waves have an unusual property: An infinite number of them can exist in the same location. If time and space are one day shown to consist of quanta, the quanta could all exist piled together in a single dimensionless point".

Let's examine this.

"An infinite number of them can exist in the same location" = True if we use a "Type 1" Infinity or "Type 3" Infinity.

"the quanta could all exist piled together in a single dimensionless point" = False. A dimensionless point permits no quanta of any kind–it is, by definition, the absence of any quantity. A point can be used to locate an object but it is not an actual volume of space.

Waves and quanta need "somewhere" to exist. Whether we are dealing with a 1-dimensional value or a 100-dimensional value, they can never be contained within a 0-dimensional point.

Singularities, Black Holes and the Big Bang

Black Holes are generally assumed to contain at their hearts a mathematical singularity–a region of infinite density. The logic of this is simple: a volume of space gets smaller and smaller due to gravity until it completely disappears. It is assumed that this zero-sized point still contains all the matter it contained before it disappeared.

This is mathematically equivalent to dividing a line in half again and again until you end up with zero. The major flaw with this is that you never reach zero. No matter how much you shrink space it will always occupy a non-zero volume. Stated another way, you can subtract volume to get to zero, but you cannot divide it all the way to zero.

Our knowledge of Black Holes has evolved over the past hundred years. There are several known limits and boundary conditions known to apply to Black Holes.

In 1930, astrophysicist Subrahmanyan Chandrasekhar calculated that a body above 1.44 solar masses would collapse under its own gravitational field. This limit became known as the Chandrasekhar limit.

In 1939, Robert Oppenheimer and others predicted that stars of more than around three solar masses would collapse into Black Holes for the same reasons shown by Chandrasekhar. This limit is known as the Tolman-Oppenheimer-Volkoff limit. This is the mass beyond which a Black Hole forms and develops an event horizon.

If the mass of the star exceeds the Tolman-Oppenheimer-Volkoff limit even the "degeneracy pressure" of its neutrons is not enough to stop a collapse. What happens during this collapse is hidden from us because not even light cannot escape.

Roger Penrose introduced the concept of the Black Hole singularity in 1965 in his interpretation of gravitation collapse, based on General Relativity Gravitational collapse and space-time singularities. John Wheeler coined the term "Black Hole" in 1967.

The assumed existence of singularities in Black Holes has been used by Stephen Hawking and others to conclude that the Universe began from a singularity. This assumes that a point with zero dimensions grew and grew until it reached the size we see today.

Essentially Hawking ran the Penrose model backwards: instead of something (matter, energy, space and time) being sucked into nothing, in the Hawking model nothing suddenly expanded into something (matter, energy, space and time).

There is still the same fundamental problem in this model as in the Penrose singularity. It assumes that merely by multiplying by zero many, many times, that you can create something. Simply put, it requires that 1×0 is greater than 0.

This is not to say that Black Holes do not exist. We make no predictions here about trapped surfaces or event horizons. We merely demonstrate that there is no such thing as a zero-sized singularity at its centre–these arise purely from flawed logic and the incorrect treatment of zero as "just another number".

If there are infinities in both Black Holes and The Big Bang they are "Type 1" Infinities (1 / Infinitesimal), not "Type 2" Infinities (1 / zero). We can get as close to zero as we wish, but never actually reach it.

The apparent "Type 2" infinity is just a mathematical error introduced when we focus on a zero-sized point in a curve. There is no need for "Cosmic Censorship", the term used to describe how nature hides the singularity from us. There is, quite literally, nothing to see in a zero-sized point.

Similarly, a spinning zero-sized volume is no different to a stationary zero-sized volume, so predictions of rotating singularities leading to worm-holes are also questionable. In this case as well, treating division by zero as an actual value is the root cause of the confusion.

It is possible that there is another limit associated with density in Black Holes. Let's call it a "Penrose limit". This is the limit at which gravitational collapse ceases or is held in check by some other force or forces. In this case the density would be finite.

The alternative to this is that the mass continues to collapse forever in an ever-decreasing volume which is always greater than zero. In this case the density at any time is still finite, although it is ever-increasing, like one of Aristotles "potential infinities".

Space and Time

As previously discussed, zero space means no space at all, not just a tiny amount. A volume of zero is out of scope for any calculation of volume as zero volume is not a volume at all: it is a point or the absence of any volume. The same holds true for time. If a point does not occupy any space, it cannot occupy any space-time (operations such as "space × time" or "space / time" return no value). If any dimension is absent, it cannot exist in a real 4-dimensional Universe.

We can imagine a cube with 1 centimetre sides but can we imagine a cube that exists for 0 time? If the x, y and z coordinates of the cube are 1, but the time coordinate is 0 then it can never actually exist in the Universe we observe.

Removing the "divide by zero" or "Type 2" infinities from space and time leaves us with the "divide by infinitesimal" or "Type 1" and the "never-ending quantity" or "Type 3" infinity.

This also demonstrates again why we so often hear "do not divide by zero". Although the explanation is seldom discussed the answer is simply that there is no number and there is no result, just the absence of any number. Note once again that viewing zero as the absence of a number rather than as a number makes no difference in addition, subtraction or multiplication, only in division. Sometimes we are not even aware that we are relying on division by zero in issues involving space and time, yet we do.

String Theory

The goal of String Theory is to describe the Universe we live in using only the simplest possible objects: 1-dimensional "strings". It is not actually a single theory but a number of related approaches based on similar fundamentals.

Imagine a one-dimensional object–a line. It has length but no height or depth. As we have previously observed in *Points, Lines, Planes and Cubes*, an infinite number of points cannot be used to construct a line, an infinite number of lines cannot be used to construct a square and an infinite number of squares cannot be used to construct a cube. Objects that have zero height, no matter how high you stack them, will never gain height.

String Theory is a theory based on division by zero. Without division by zero and "Type 2" infinites it cannot exist. Unless we completely redefine what a dimension is, String Theory falls at this hurdle.

This is because of the assumption of vibrating 1-dimensional objects being at the root of all we observe in our multi-dimensional Universe. A 1-dimensional object can only vibrate in one dimension. Any meaningful motion in any other dimension is limited by the fact the object lacks any existence.

$$\longleftrightarrow$$

Vibration in single dimension

To make matters worse, in String Theory a 1-dimensional object can form a loop. The thinking behind this is that a line has only one dimension (length) and that by joining the two ends together we still have a 1-dimensional object. In our multi-dimensional

Universe, however, a loop requires 2 dimensions: length and height.

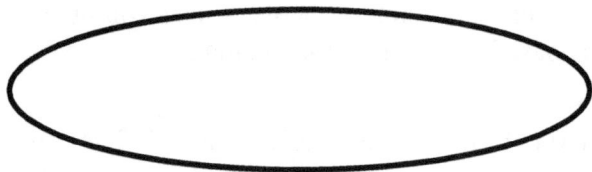

Looped string

And even this only gives us a 2-dimensional object, which is at least 2 short of the accepted 4, without worrying about 10, 11 or 26 dimensions.

Remember that the lines in the diagrams above are purely to conceptually represent the objects. The actual objects are 1-dimensional and have no size at all where we have shown a line.

Another way of looking at this is that if an object has a zero value in a particular dimension, then it simply does not exist in that dimension. Single dimensional objects can be used to measure the actual Universe we live in, but they cannot be used to construct it.

The goals of String Theory are admirable and it is a theory that needed to be looked at, but String Theory has occupied a generation of physicists without ever producing that "Eureka!" moment and it is unlikely that it ever will while we live in a 4-dimensional Universe.

Renormalization

Renormalization involves removing infinities from the results produced by a theory. These infinities fall into a number of categories.

a) An infinite number of paths.

Cases where as an electron or photon is moving from A to B.

It may take one of an infinite number of paths.

This is a "Type 1" infinity.

b) Divisions by zero.

Such as $x = y / r$ where $r = 0$, a zero radius.

This is a "Type 2" infinity.

c) Infinite number of terms.

Cases where integrals are divergent, that is, they give infinite answers. For example, when assuming that fields have values at an infinite number of points in space.

These are "Type 1" or "Type 3" infinities.

d) Infinite distances.

For example one object may be up to an infinite distance from another.

This is a "Type 3" infinity.

e) Forces acting over infinite distances.

For example the effect of one object on another, up to an infinite distance away.

This may involve both a "Type 3" infinity of distance and, possibly, a "Type 1" infinity for force.

Some of the infinities in areas such as quantum field theory arise not because of "Type 2" infinities (division by zero infinities, which do not actually exist), but from "Type 1" infinities (division by non-zero value).

Whether these infinities in field theory are truly infinite or merely very large depends on whether space and time have a granularity at very small scales, whether there is a smallest possible distance or shortest time. According to current theory, this may be the Planck length and the Planck time, although this is speculative.

It may simply be that the current quantum field theory is just not quantised enough. As we have discussed earlier, after we eliminate type the "Type 2" (division by zero) infinities, the only infinites that can apply to field theories are "Type 1" infinities (energy, mass, space or time divided into small sections, not zero-sized points) and "Type 3" infinities (unlimited forces, times or distances). This leads to the possibility that many of the infinities found in field theories may perhaps not be infinities at all, but finite values.

8 Zero and Computer Software

The use of zero in computer software is often used to illustrate the "weirdness" of zero. However, the actual truth is much more straightforward.

Zero and Nothing In Software

As mentioned previously, the difference between Number Zero and Nothing is best illustrated using the set theory. The conventional view of "number zero" is { 0 }, whereas "nothing" is { } or the empty set.

Computers are quite capable of distinguishing between numbers and nothing. In computing, the absence of any value is called "Null" or "Nothing". Software also often uses zero as a special value to indicate a "null pointer", or a pointer with no value that doesn't point to anything. This is treated as a special case–as the absence of a pointer, not as an actual pointer.

Division By Zero In Computer Software

The case of division by zero in computer software is sometimes raised to demonstrate the slippery nature of zero. In fact, raising an error when division by zero is attempted is an action that is deliberately programmed into computers on the assumption that it is not possible. There is nothing innately in them that cannot allow them to take a different action.

Problems with division by zero only arise when the error is not handled correctly; this is called an "unhandled exception".

Similar unhandled exception conditions could include not finding a file that the program needs or running out of memory. Any unhandled exception, whether division by zero or running out of memory, will cause problems for the computer. There is nothing inherently difficult about zero that causes a computer any problems; it just does what it is told.

In software division by zero is generally avoided by treating it as the absence of any value and not attempting it. If a program attempts it, it will find another piece of software or hardware telling it that it should not have tried. If hardware manufacturers change their arithmetic processing units to simply return zero instead of generating an error; and the creators of programming languages also treat the result of division by zero as zero then the problem would not exist.

It is equally correct to simply set a limit at zero and to never attempt the division. Given that zero is the absence of a number, rather than a number in its own right, the result of this division is zero and this should be treated as "nothing".

It is well known that attempting division by zero in computer software will cause an error. What is often overlooked is that the correct course of action is to discard the value.

There is no information content in the result–that is why computer programmers discard it. If zero is a number then we would not and should not discard it. The fact that we can and do ignore the results of division by zero is that zero is not a number and is outside the scope of the equation.

Issues of Scope In Computing

There is generally a difference in the way that computer programmers and physicists view equations. In software development, the "scope" of an equation is always a matter of concern.

If a theoretical physicist makes a mistake with the scope of a function, then in a year, or a decade or a century, someone will publish a correction. If a computer programmer makes a mistake in the scope of a function, something goes bang, someone loses money or planes fall from the sky.

Division by zero is generally avoided by setting a limit at zero, although the reasoning behind this is often overlooked. Division by zero can and should be avoided because there is no number to divide by and if you do divide by zero, you get no number as a result.

Endless Loops and Counting To Infinity

Another type of infinity regularly encountered in computing are "Endless Loops". These are cases where a series of instructions is repeated endlessly, never reaching an "end condition". This is similar to our "Type 3" Infinity. Some programs are designed to run endlessly–heart monitors for example. Other activities, like counting cats or a game of chess, are usually assumed to have a finite execution time.

Note that no computer has managed to count to infinity while in an endless loop–the program is usually terminated after seconds, minutes hours or days, months or years. Fifty years would be a very long lifetime for an endless loop using current technology. So although it is a good lifespan for any man-made machine, it is well short of infinite.

85

The usual limit encountered when computers count, such as counting cats for example, is the storage required to keep the count. If we wish to count up to 65536 cats then 16 binary digits are sufficient. If you wish to count up to 4 billion then 32 bits are required. For 4,722,366,482,869,650,000,000 cats a mere 72 bits of information are required.

Most computers produced today could represent astronomically large numbers with at least 100 billion bits. In theory we could one day count the number of protons in our Galaxy, given enough time and enough bits of memory to store the count. This is still well short of infinite however.

In actual practice in computing, infinity is never directly represented, typically just a single bit is used to indicate whether a number is "infinity" or not and to treat it as a special case if it is.

9 The Cult of Zero

Everyone loves a mystery, even mathematicians and physicists, especially *mathematicians and physicists. There exists in some circles an almost mystical approach to zero and infinity. Much of this harks back to the Pythagoreans, the "infinite void" and other philosophies. Some people simply choose to revel in the mysterious zero-infinity.*

Hearts and Minds

Even today we find that the mystical singularity in Black Holes attracts all kinds of ideas on what happens when "the laws of physics break down". Such is the degree of belief in "number zero" that some people would rather believe that the laws of physics break down inexplicably than that zero is, at least sometimes, not a number.

Modern mathematicians sometimes write condescendingly of the Indian Brahmagupta who recorded in the seventh century that the result of zero divided by zero was zero. The "sophisticated" modern mathematician believes that zero is always a number and that division of zero by zero returns a value called "undefined" (which doesn't actually exist as a number). Similarly, many writers on the subject belittle those who rejected the idea of zero as a number.

This smugness is very misplaced. Brahmagupta and other ancient mathematicians saw some things far more clearly than we do. It has taken the confusion of modern mathematics to muddy the perfectly simple idea that we can have a complete absence of something.

As it turns out, those in earlier times had rightly voiced their discomfort over the problems created by "number zero". Their concerns were based on solid ground and those who rejected that any number divided by zero must be infinity or undefined were absolutely right to do so.

Unfortunately, such is human nature that some people, even when presented with overwhelming evidence, stick doggedly to their beliefs, rational or not. Some ideas, like Evolution or the Sun-centric solar system, can take generations to be accepted.

10 Conclusion

We have had a brief introduction to the history of the many things we call zero. We have examined the realities of geometry and physics based on how things really are rather than on the self-referential system of conventional mathematics. What's next?

Changing Paradigms

In this book we have examined just a few of the troubles with zero. There are undoubtedly many more in areas from mathematics to geometry to physics.

The hallmarks of a good theory are that it should be self-consistent and consistent with the world we observe. The conventional views of zero as a number fail on both counts. It requires "undefined" operations and "impossible" results and does not fit with the mathematics of the Universe we observe.

Human beings are clever enough to send a man to the moon and to estimate the weight of a nearby galaxy yet we cling to mathematical conventions that clearly have problems.

When we dogmatically cling to a bad idea, as with the concept of an Earth-centric universe for example, errors and contradictions arise which are then suppressed or dismissed. This occurs when we hold that zero is always a number and that there is some type of almost mystical relationship between zero and infinity. We find we have to introduce illogical concepts to make our cherished idea work.

Just as the Earth-centric universe demanded that "epicycles" be included in order to make sense of the apparent backward motion of the planets, the "number zero" view of zero requires illogical rules such as that 1 / 0 is "impossible" or that 0 / 0 is "undefined". And, as with the discarding of the Earth-centric solar system and its replacement with our current Sun-centric solar system, it is time to abandon our view of zero as always being a number.

Abandoning these long-held views about zero will not happen quickly though. Too many teachers and academics have "drunk the Kool-Aid" and will not readily deny beliefs they have held or taught for many years. Surprisingly few academics will loudly voice an opinion that could be damaging to their careers or which could attract the scorn of colleagues. For many people it is simply easier to stick with a bad idea than to change their minds.

Regardless of this, some people will apply the principles outlined in this book and find that they work each and every time without exception and they work without the need to introduce illogical concepts or impossible results. It is for these people that this book was written.

Index